The white pyramid

Science and Research

English Edition

1. The author reports

   Pyramid - white - Marble - within 80 Km / h converts a computer all in melodic sounds around ... around the clock - one on Computerized dust and dirt on ELECTROMAGNETIC BASE - causes is dirt, dust and the like always immediately dispose of. Crystal glass - top stands out in good weather. When you enter the pyramid they seem to be completely empty - a computer adapts to all of what you need. Partly also be given visual stimuli media again if you only THINK THE FACT: In the lower part of the

property are COLLECTIONS -
under   3
including the largest collection of
musical instruments. Bottom: 6cm
Midnight Blue Flooring - Synthetic -
In the laboratory, you can make
your look with an injection for 100
years to determine the long term.
It is for all staff (Cumputer) - we
see the sober - Here is just women
and   the   ICH.
D. L. F

2. The authors

3. Research of Par1 The 4 Cuties - freundinnen- the wife says:

The love Cuties Cute mean in English sweet, cute - Cuties: Baby, Funny Face and these are the girls too. Confident they present themselves in the pictures, pornography, they play no offers itself. Cuddly. They are happy that's real, and broadcasting works through the images - sexually arousing, I'm not a lesbian - married coupon. For sexual arousal are porn. Movies anguish, I will never watch people having sex -the pain pierces my soul - Only Girls! Dirk and I have been married 12 years all started with Lindsay Marie. Suddenly I built eight sites with images of a woman unknown to me, but I was loving and attractive. Then we saw a strip-tease video with music by Lindsay, which is always nice 31. Lindsay dances for us. And Dirk and I have 3000 images so far of girls, women - and Cuties, I noticed only for a few months. This is ideal for couples, from a psychological point of view - I was not jealous of women who are "better" than I looked - I am one of them. Field research, statistics, study - a help to many couples - even speak openly so and then livecam because it's different. Since I was a few times angry, angry. But I myself had the idea of Free Live to see - the stress drilled in our neck of writing these lines at all the books are a luxury for me. When I'm in the Books read about stress, I can only shake his head about what I had been upset before, got a bad mood. It began on Mother's Day. Dirk and I do not have time to analyze the long time because it's so crazy, but real. But now to the Cuties, of course I have compiled collection.

4. Global change - from global reform - the author says:

Global Reform - Statement 10/12/2014 talking about the system from I I've tested in 1989, it works in the same way, the same principle is already 100% in Australia to the slope of the total power (Opens in gamers not conscious areas) I've sold the report skimmed over, it is around $ 250.00 - sold. As soon as I find it again I share it with you. Non psychiatrist. My system I was ausbaue further since 1986 in phases 2 - 6 years tested. Tested on myself, then a satellite - initially to Troubleshoot Your Own ICH, with every thought, "What could I my   13 Fellow Antun ", electrical shock, uncomfortable at first, then later desired. (Simple Design) Originally the idea - the idea of fighting crime - Target harmony in all societal layers, since my family only logical thinking - my own idea was tried out on me. After this cleaning, I did not understand until some time do not why people commit crimes - sometimes it was so bad that I had to vomit - 1 - 2 times because people in my immediate area who spoke did over Greul - particularly in the area of what people all have done with women, the lack of or no respect, before families - not   requested   obvious PENETRATION in private taboos, I refer now only in the intellectual, monitoring - tests by unsubstantiated rights which have been made of sizes - or do not own knowledge-based, which was used instead of the normal to share his brother or his sister to use it in a way - which had to be clarified in terms of intelligence since 1986 for this planet. Seen from my point of view - this kind of approach I personally think is the cause of the "very severe disease" catatonia. A few short phrases: We live in a world where no one has to die, who will not. Just for you I imagine it must have exactly understand why we do not use our technology has long been true.   15

About Mr. Sorbo, I want to tell you
at this point, with respect. Thank
you your message. There is a lot to
do, as you can see from this little
report out. A few short phrases:
We live in a world where no one
has to die, who will not. Just for
you I imagine it must have exactly
understand why we do not use
our technology has long been
true. There are no more middle
class - but more and more people
are discovering themselves -
starting to write the androids are
our insurance, many others, I have
determined many times over the
machines that we have today,
such as to laugh our iPhone. On
the one hand I say, for example,
times when my wife was asleep,
You to me are not going well, can

you help me somehow, they said, let me "think" they said, I know a joke right on: "Meet two iPhone's. ? Helps the "My wife asked Siri, she had her IPhone just getting started, I must be lying because Siri replied:" That can not be, "My wife has 14 days not spoken with the iPhone.. We must always remember that we are creating these machines! So if you are only able to respond so you only can gather from the internet, what we put in knowledge if we irresponsible deal with it, we feel it yourself, we feel ourselves self, we experience ourselves, then we are GOD - a unit and so only the death will cease, and we will finally expand into the universe and explore new planets. With everything else is now easy closing. Done. 17 There is a GOD. You even say, he must prove it. There is nothing left to prove, all right? A man says this here today - how can that be, what should the questioning at all, you have the God-conscious - they would have done it. Thank you. (2 weeks ago - I was simply sterilized so, or is it due to a finger needs to be sewn in sick - supervised by an 8-headed Medical team - 2 policemen keep watch, you know why I do not do anything in this matter - Someone, I have therefore gebeten- for a person to believe me, did the same with my wife did, but only because she wanted it that way. I think it's

Bold and fun of my fellow human beings, but only because I have my eye destroyed and repaired itself   .)   I see you right something else - something that can WILL POWER. Only 2 images. Dirk L. Feiler Supreme Commander of the US president Dirk L. Feiler Why do I have this position for so long - I am clearly IRRE - if you mean - not true - I just did not have enough time to educate myself. Nevertheless, says Barack good, it is not without you, sorry Barack - silly game I know! Do you want a miracle? But if you want that then I would be very, very nice to me in your place - somehow the

United States is compared with, for example, Google helpless? Or something like that. The planet is a small world full of children. I know God - for this being we are something like 1 cent ... tip. Notice what they do, that people read also formed. Education that you do not make a right use ... llol - I'Sie all Dear! Caution ISIS is now no longer in: From now on, miracles will prevail. Sounds nice IS is I eat so much and when I want 65% of people agree with me. Or want Sun 0%. Sorry, I'm just mad. Ask Kevin Sorbo. Maybe he tells them something that takes! Why do I have this position for so long - I am clearly IRRE - if you mean - not true - I just did not have enough time to educate myself. Nevertheless, says Barack good, it is not without you, sorry Barack silly game I know! Do you want a miracle? But if you want that then I would be very, very nice to me in your place - somehow the United States is compared with, for example, Google helpless? Or something like that. The planet is a small world full of children. I know God - for this being we are something like 1 Cent ... tip. Notice what they do, that people read also formed. Education that you do not make a right use ... llol - I'Sie all love! Caution ISIS is now no longer in: From now on, miracles will prevail. Sounds nice IS, I is eat so much and when I want 65% of people

agree with me. Or want Sun 0%.Sorry, I'm just mad. Ask Kevin Sorbo. Maybe he tells them something that takes! Sorry education should be !!!
I am also a "man." A little knowledge never hurts - but without government - what you do - damn it - it's so simple - Never again someone will have to work if he does not - it's not necessary - you study yourself why I have this position already so. long - clearly I'm IRRE - if you mean - not true - I had only 21 not enough time to educate myself. Nevertheless, says Barack good, it is not without you, sorry Barack silly game I know! Do you want a miracle? But if you want that then I would be very, very nice to me in your place -

somehow the United States is compared with, for example, Google helpless? Or something like that. The planet is a small world full of children. I know God - for this being we are something like 1 cent ... tip. Notice what they do, that people read also formed. Education that you do not make a right use ... llol - I'Sie all love! Caution ISIS is now no longer in: From now on, miracles will prevail. Sounds nice IS, I is eat so much and when I want 65% of people agree with me. Or want Sun 0%. Sorry, I'm just mad. Ask Kevin Sorbo.   Perhaps he tells them something that takes! Sorry education should be !!! I am also a "man." A little knowledge never hurts - but without government - what you do - damn it - it's so simple - Never again someone will have to work if he does not - it's not necessary - you do not study enough even your mind to see that. - You prefer death - wars and so a nonsense - I offer them to all the real paradise, you are all stars. OMG Why do I have this position for so long - I am clearly IRRE - if you mean - not true - I just did not have enough time to educate myself. Nevertheless, says Barack good, it is not without you, sorry Barack - silly game I know! Do you want a miracle? But if you want that then I would be very, 23 very nice to me in your place somehow the United States is compared with, for example, Google helpless? Or something like that. The planet is a small world full of children. I know God - for this being we are something like 1 cent ... tip. Notice what they do, that people read also formed. Education that you do not make a right use ... llol - I'Sie all love! Caution ISIS is now no longer in: From now on, miracles will prevail. Sounds nice IS, is - I eat so much and when I do 65% of people agree with me. Or want Sun 0%. Sorry, I'm just mad. Ask Kevin Sorbo. Maybe he tells them something that takes! Sorry education should be !!! I am also a "man." A little knowledge never hurts - but without government - what you do - damn it - it's so simple - Never again will someone have to work if he does not - it's not necessary - you do not study enough even your mind to recognize that. - wish you would prefer death - wars and so a nonsense - I offer them to all the real paradise, you are all stars. OMG But not from a IDIOTS GOVERNMENT - we can do it all yourself what you believe to think ... you think NOTHING - not even the stars are the worst off. Man sin you stupid. Brew today much or little government - just say - they are ridiculed look ha, ha, ha The last President Barack Obama to get the house and then we humans think without a government.. Well, not then. There are funeral expenses insurance! System, the androids, the guardians of our SELF - It is the only possible androids collected all Human knowledge "to use against us" (now better understand) The right approach our knowledge will expand and indeed as far as the machines are commonplace - Never again will create knowledge that we ourselves use against us by our androids to us , This is the design of a truly TRUE global FRIEDENSGARANTI E. First plan was anyone to do what you want in no case itself, "3 Years experiment on himself" all ABSCHSCHEULICH ill do what urging me on and temptation such thoughts even "unconsciously take over satellite with a kind of power (painful) to stop immediately) worked in every level of society. prototype of the 80 years. There is no middle class more - but more and more people are discovering themselves - starting to write the androids are our insurance, many others, I have determined many times over the machines that we have today, such as to laugh our iPhone. On the one hand I say, for example, times when my wife was asleep, You to me are not going well, can you help me somehow, she said, let me "think" they said, I know a joke right on: "Meet two iPhone's. ? Helps the "My wife asked Siri, she had her IPhone just getting started, I must be lying because

Siri replied:" That can not be, "My wife has 14 days not spoken with the iPhone.. We must always remember that we are creating these machines! So if you are only able to respond so you will only read out from the Internet 27 to what we put in knowledge, irresponsible if we deal with it, we feel it yourself, we feel ourselves self, we experience ourselves, then we are GOD - a unit and so only the death will cease, and we will finally expand into the universe and explore new planets. With everything else is now easy closing. Done. There is a GOD. Laid eggs at the very beginning - as dinosaurs. / The seventh in this system, the last - "everybody knows." In I think in the 1600 century I have for the ladies of the court where ever played harp. War has not allowed me father. And I also sing other languages 99% correct. Spontaneous. It often takes years to understand the meaning. And I can because "code" also share how you too. So far, only in singing. Unfortunately, not with my wife. I am glad that you were able to sleep. Okay, then we will prepare ourselves for the new WONDER - no we do not do what you exactly want! Pleas E Ask Your Question! A few short phrases: We live in a world where no one has to die, who will not. Just for you I imagine it must have exactly understand why we do not use

our technology has long been
true.

...   To   be   continued